Housing
231

你迷路了吗？

Are You Lost?

Gunter Pauli

[比] 冈特·鲍利 著

[哥伦] 凯瑟琳娜·巴赫 绘

李原原 译

上海远东出版社

丛书编委会

目录

你迷路了吗?   4

你知道吗?   22

想一想   26

自己动手!   27

学科知识   28

情感智慧   29

艺术   29

思维拓展   30

动手能力   30

故事灵感来自   31

Contents

Are You Lost?   4

Did You Know?   22

Think about It   26

Do It Yourself!   27

Academic Knowledge   28

Emotional Intelligence   29

The Arts   29

Systems: Making the Connections   30

Capacity to Implement   30

This Fable Is Inspired by   31

食蜂鸟和褐胸鹟一边聊天，一边为它们下一次到世界另一头的长途旅行做准备。

"嗯，我准备飞到印度去。"褐胸鹟宣布。

A bee-eater and a flycatcher are chatting while preparing for their next long trip to the other side of the world.

"Well, I am just about ready to fly off to India," the Brown-breasted Flycatcher announces.

食蜂鸟和褐胸鹟在聊天……

A bee-eater and a flycatcher are chatting ...

……孩子们的第一次旅行。

... first trip for our youngsters.

"太好了，你要去印度，而我要往南飞，回到我的祖国澳大利亚。"蓝尾食蜂鸟回应道。

"这将是孩子们的第一次旅行。它们以前从未迁徙过，但我相信他们会朝着正确的方向起飞。"

"Great, you're off to India, and I'm flying down south, back to my native Australia," the Blue-tailed Bee-eater responds.

"It will be the first trip for our youngsters. They have never migrated before but I'm sure that they will take off in the right direction."

"是的，我们都必须学会使用我们的视觉、嗅觉、听觉，当然，还有我们神奇的磁传感器。"

　　"人类一直想知道我们是如何做到如此精确导航的。还记得他们是如何把金属环戴在我们的腿上，以弄清我们是从哪里来的吗？现在他们使用数学和生物学来实现这一目的。"

"Yes, we've all had to learn to use our sense of sight, smell, hearing and, of course, our magical magnetism sensors."

"People have always wondered how we could navigate with such precision. Remember how they used to put metal rings on our legs to figure out where we came from? These days they use mathematics and biology."

我们神奇的磁传感器……

Our magical magnetism sensors ...

我们喙里的铁，以及我们眼睛里的光传感器……

Iron in our beaks and light sensors in our eyes …

"经过一个多世纪的研究，他们现在才弄清我们是如何找到一条安全的直接到达目的地的路线！他们在我们的喙里发现了铁，在我们的眼睛里发现了光传感器，但他们仍然无法弄清……"

　　"现在，如果要让我研究我们的机能和成功秘诀，我宁愿从我们的超级食物入手。" 褐胸鹟说。

"Over a century of research, and they are only now figuring out how we find a safe route directly to our destination! They found iron in our beaks and light sensors in our eyes, but still they could not explain …"

"Now, if I were to study our performance and success, I would rather look into our superfoods," Flycatcher says.

"嗯，我们紧紧锁住脂肪，为飞行提供能量。我们通过减轻体重来提高效率。我们向血液输送氧气，我们的肌肉不仅用于飞行，还能用于加速。"

"我们还寻找积云。一旦我们被这些蓬松的东西包围，我们就可以闭上眼睛，打个盹，享受强烈的上升气流。"

"谢天谢地，人类现在开始关心我们的生存状况了。"

"Well, we do lock up fat, to power our mission. We reduce our body weight to increase efficiency. We oxygenate our blood, and use muscle tissue not just to fly, but to turbocharge our flight."

"And we will look for cumulus clouds. Once we're around this fluffy puffy stuff, we are even able to close our eyes, take a nap and enjoy the strong updrafts."

"I am so grateful that people nowadays care about our survival."

······我们寻找积云。

... we will look for cumulus clouds.

......人类怎么就喜欢射杀鸟类......

... how people can shoot at birds ...

"是的，我永远也搞不懂人类怎么就喜欢射杀鸟类，还试图刷新日猎鸟记录。他们娱乐时为什么就不能只射击泥鸽呢？"

"幸运的是，时代在变，人类的环保意识也在增强。现在他们要求把我们最喜欢的着陆点，也就是我们休息和获取食物的地方，专门保护起来。"

"Yes, I will never understand how people can shoot at birds, trying to set a record with the number of birds killed in a day. Why can't they just stick to shooting clay pigeons for their sport?"

"Fortunately, times are changing, and people are more aware. They are now asking that our favourite landing spots, where we rest and get some food, are kept just for us."

"更可喜的是，在一些地方，人类正在重建湿地，让汽车和游客远离岛屿，以便我们可以自由通行。真令人感动！"

"的确让人感动。"食蜂鸟表示赞同。"让大自然重生，帮助我们安全抵达繁殖地……"

"但我们现在又面临一个新的危胁，就是人们的最新嗜好：手机。"

"Even better, in some places people are rebuilding wetlands, and keeping cars and people off islands for us to have free passage. Impressive!"

"Impressive indeed," Bee-eater agrees. "Regenerating Nature to help us safely reach our breeding grounds…"

"But a real danger we are faced with now, is people's latest addiction: their phones."

重建湿地……

Rebuilding wetlands ...

……世界各地的天线。

... antennae everywhere.

"我知道，不仅仅是手机，还有他们装在世界各地的不计其数的天线。借助这些天线，他们就能通过手机更多更快地往全球发送信息。"

"人类现在应该明白，这样对我们不好，对昆虫和植物也不好。"

"I know, and it is not just the phones, but also the thousands, even millions of antennae they are putting up everywhere to send more information around the globe faster, through their handheld devices."

"Surely people should know by now that it is not good for us, and not good for insects and plants either."

"完全正确！难道他们不明白过度总是不好的吗？"

"嗯，看来只能让他们从痛苦中吸取教训。一开始，他们会感到困惑，然后变得愤怒。不幸的是，只有当环境受到损害时，他们才会愿意改变，并最终改变自己的生活方式！"

……这仅仅是开始！……

"Exactly! Don't they know that an excess of anything is always bad?"

"Well, it seems they only learn the hard way. At first, they will be confused, and then they become angry. Unfortunately, it is only once the environment gets damaged, that they will embrace change, and finally change their ways!"

... AND IT HAS ONLY JUST BEGUN!...

... AND IT HAS ONLY JUST BEGUN! ...

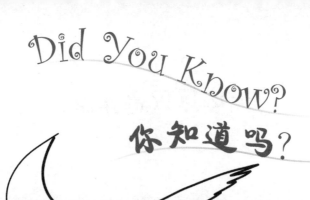

The Arctic tern flies approximately 2.4 million kilometres in its lifetime. This equals flying to the moon and back 3 times. While a direct line between two places may be 400,000 km, the tern will fly up to 100,000 km per trip.

北极燕鸥一生大约飞行 240 万千米。这相当于从地球到月球往返 3 次。地球和月球之间的直线距离大约 40 万千米，而燕鸥每次飞行的距离可达 10 万千米。

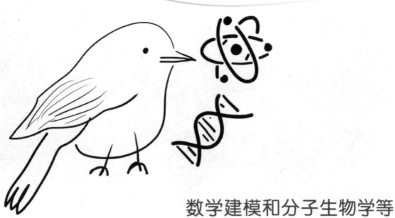

Modern science, including mathematical modelling and molecular biology, offers more insights than a century of ringing birds.

数学建模和分子生物学等现代科学的出现，使得如今人们对鸣禽的理解远超过去一个世纪的研究成果。

候鸟是耐力运动员。它们通过吃富含 Ω-3 脂肪酸的食物来增加肌肉的供氧量。如果没有这种超级食物，鸟类将无法完成飞行。

Migratory birds are endurance athletes. They increase the amount of oxygen available to muscles by eating Omega-3-rich food. Without access to this superfood, birds would not be able to complete the trip.

鸟类结合声音、嗅觉、视觉和磁场来找到正确的方向。幼鸟在晚上起飞，开始它们的第一次旅程。它们会选择正确的方向，不会有任何错误。

Birds use a combination of sound, smell and sight, along with their magnetic sense, to find the right direction. Young birds taking off at night for their first journey will choose the right direction without any error.

The greatest threat to migratory birds is the annual slaughter of an estimated 25 million illegally hunted birds. The birds are able to escape hurricanes and typhoons but not the hunters' bullets and birdshot.

每年大约有 2 500 万只候鸟被非法猎杀，这是对鸟类最大的威胁。这些候鸟能躲过飓风和台风，但躲不过猎人的子弹和鸟枪。

As a result of habitat destruction, migrating birds' favourite stop-overs are disappearing from one year to the next, forcing them to continue on without food, water or rest, making them very vulnerable.

由于栖息地被破坏，候鸟喜爱的中途停留地每年都在消失，迫使它们在没有食物、水和休息的情况下继续迁徙，这使它们变得非常脆弱。

Electromagnetic radiation from power lines, Wi-Fi, and phone masts, pose a credible threat to wildlife. New communication technologies like 5-G could cause even greater harm.

来自输电线、Wi-Fi 和信号塔的电磁辐射对野生动物构成了实实在在的威胁。像 5G 这样的新通信技术可能会对这些动物造成更大的危害。

The EU-funded EKLIPSE project, that studies ecosystems and biodiversity, found that electromagnetic radiation not only poses a risk to birds, but also to insects and plants. To date, there is no definition of a safe limit of electromagnetic radiation.

由欧盟资助的 EKLIPSE 项目研究生态系统和生物多样性。科学家通过该项目发现，电磁辐射不仅对鸟类造成威胁，而且对昆虫和植物也有威胁。迄今为止，还没有对电磁辐射的安全极限的定义。

Would you protect wetlands and keep people out, to keep birds migrating?

你会为了让鸟类迁徙而保护湿地，并因此不让人类进入湿地吗？

Can you imagine people shooting birds for sport?

你能想象人们为了娱乐而射杀鸟类吗？

Would you lose body weight to be more efficient, or gain weight to have reserves?

你会为提高效率而减肥，还是会为储备力量而增重？

Is it always necessary to learn the hard way?

有必要总是通过痛苦的方式来学习吗？

Do It Yourself!

自己动手!

Do you have any idea which birds fly over your home when migrating? Do you live in or near an area where migrating birds take a rest? Find out which birds are using the airspace above your home, or a piece of land in your area when migrating. When you know which migratory birds these are, and more about the risks they suffer, and you are clear that these birds merit your support, draft a plan to help protect them. Share your findings and your plan with friends and family, and gain their support.

你知道有哪些候鸟在迁徙时飞过你家吗? 你是否生活在候鸟的停歇地附近? 找出那些迁徙时在你家附近停留过的鸟类。当你知道这些是什么候鸟, 更多地了解它们所承受的风险, 并且清楚这些鸟值得你的帮助时, 起草一份计划去保护它们。与朋友和家人分享你的发现和计划, 并获得他们的支持。

学科知识
Academic Knowledge

生物学	为了准备迁徙，鸟类进入"暴食"状态，大量进食以储存脂肪，作为迁徙的能量；有些鸟，比如黑顶白颊林莺，在连续飞行86小时、3 700千米之前，体重几乎翻了一番；大脑中的磁传感器。
化 学	脂肪合成，为迁徙积累能量；迁徙所需要的能量主要通过脂肪酸的氧化反应提供；脂肪组织中存在富含能量的甘油三酯(TAG)；脂肪酸碳链的长度、不饱和程度和双键的位置都会影响飞行中储备脂质的氧化速率；血液氧合的过程。
物 理	鸟巢具有不对称的弹力结构；鸟喙中的金属和眼睛中的光传感器可以指引鸟类迁徙；鸟类利用上升气流；过强的无线电波可能影响鸟类迁徙。
工程学	通过了解生物学和预测迁徙路径，确保地理定位和运动监测。
经济学	节约和建立储备，从而有能力采取重大举措的重要性；一方面利用一个地区的过剩产品，另一方面规避其他地区的短缺。
伦理学	每年有数以亿计的鸟类死于撞击窗户，数以百万计的鸟类死于撞击电视和广播信号发射塔；每年都有数以百万计的鸟类死于猎鸟者的手中，这种猎杀只是为了取乐。
历 史	在古希腊，雅典娜的圣鸟代表重生；在埃及法老时代，猎鹰被认为拥有守护的力量，并与皇室联系在一起；18世纪的人们相信燕子在初秋的时候沉入泥沼，在第二年春天又以两栖动物的形式出现。
地 理	积云的形成。
数 学	利用数学方法，求出鸟巢的最大密度，或稳定状态下的密度。
生活方式	即使是不会飞的鸟也会迁徙，鸸鹋是一种大型的澳大利亚鸟类，经常步行数千米寻找食物，许多企鹅通过游泳迁徙；十月是发现有翅膀的旅行者的最佳时间；超级食物的出现。
社会学	飞翔的鸟直接与轻盈、自由等概念联系在一起，因此有了"像鸟一样自由"的表述；随着季节变化而迁徙的候鸟群排成典型的V形队列。
心理学	梦见鸟体现了幻想、理想和思想；当新的事实出现时，我们先是困惑，然后是嘲笑，接着是攻击，最后是拥抱变革。
系统论	随着季节变化，鸟类从资源缺乏或正在减少的区域迁徙到资源（食物和筑巢地点）丰富或不断增加的区域；保护候鸟的中转地及重建湿地。

情感智慧
Emotional Intelligence

褐胸鹟

褐胸鹟在谈话中表现出自己的同情和自信。她谈到了鸟类的生活，人类的存在，以及鸟类与人类的联系。她强调了技术的进步，但没有对此发表任何明确的观点。她引导话题，并准备分享褐胸鹟利用积云的智慧。她对猎鸟行为表示愤慨。她赞同时代在变化，并感激人类在保护候鸟迁徙通道方面做出的积极努力。然而，她非常关注电磁辐射，尽管她无法解释其中的原理，但她警告人类应该警惕电磁辐射可能产生的负面影响。

食蜂鸟

食蜂鸟也加入了友好的谈话。起初，他并没有提出任何新的见解，只是证实一些大家都知道的内容。他相当怀疑人类的理解能力，对于人类解释候鸟是如何航行的就更不用说了。不过，后来他对鸟类肌肉的新陈代谢提出了一些见解，这是一个有用的新想法。食蜂鸟对人类的关注点从解释候鸟如何成功完成迁徙转为担心候鸟的生存感到惊讶。鼓舞他的是，人们不仅研究鸟类，而且还积极保护和重建湿地。他加入了对无线电波的思考，并想知道当数以百万计的发射塔被使用后会发生什么。食蜂鸟补充了一个常识：任何东西过量都是有害的。

艺术
The Arts

鱼群和鸟群有一种魔力。它看起来好像是一个有机体，但它却是由同一物种的数千个个体成员组成。幸运的是，你可以随时准备好相机，捕捉成千上万的候鸟在天空形成的画面。试着用你的艺术捕捉这种神奇的运动，你可以在纸上或画布上画很多鸟，在运动过程中，许多单独的鸟组成一个整体——鸟群。

思维拓展
Systems: Making the Connections

候鸟从一个大陆迁徙到另一个大陆，适度地利用每个大陆的可用资源。在出发前，它们会吃昆虫来增肥。因此，鸟类有助于控制昆虫的数量。某些栖息地是鸟类理想的补给地点，对它们的生存至关重要。这些补给点一旦遭到破坏，候鸟就无法获得相应的营养。有趣的是，人们观察到，一般来说，候鸟都不去热带雨林，因为那里的食物产量是恒定的，而且没有多余的食物。这些鸟更愿意聚集在像大草原这样的地方，在那里食物的产量会随着季节的循环而变化。候鸟能够长距离飞行，它们的形态和生理机能使它们能够利用自然的力量，从世界的一个地方飞到另一个地方。候鸟在迁徙过程中需要竭尽全力，体重可能会减轻近一半。对于候鸟来说，恶劣的天气等自然现象只是它们所面临危险的一小部分。鸟类面临的威胁因不同地区的人的生活习惯而不同。高压电线和风力发电机持续对候鸟的数量造成影响。当电线或转动的发电机位于成千上万的鸟群的迁徙路径上时，就会对鸟类产生巨大的影响，使它们面临被电击或碰撞致死的危险。在长途飞行中，这些鸟类跨越许多国界，进入各种不同的地区，每个地区都有自己的环境政策、立法和保护措施。没有国际合作，任何一个国家为解决鸟类生存威胁而采取的任何措施都不会成功。某些在一个国家被认为是不可持续的做法在另一个国家可能是可以接受的。现有的一些国际协定为国际合作提供了必要的法律框架和协调文书。这些政府间条约致力于保护迁徙水鸟及其栖息地。

动手能力
Capacity to Implement

湿地有可能重建吗？候鸟是我们生态系统的一部分，它们提高了我们的生活质量，所以为鸟类留出休息、进食和取水的空间很重要。最简单的迎接候鸟的方法就是为它们预留一块不允许人类进入的地方。你家附近存在这样的空间吗？找到一个合适的区域，确保那里有充足的水源以及足够的植被。接下来，与他人分享你的想法，并一起制定行动计划。

故事灵感来自
This Fable Is Inspired by

雷切尔·穆海姆
Rachel Muheim

雷切尔·穆海姆从小就对自然和鸟类感兴趣。她在苏黎世大学学习生物学。获得学士学位后，她继续在同一所大学攻读硕士学位，研究飞越阿尔卑斯山的候鸟。2004 年，她转到瑞典隆德大学，在那里攻读动物生态学博士学位。她现在是隆德大学理学院的副教授。她对动物的磁场定位产生了浓厚的兴趣，并在弗吉尼亚理工大学的约翰·菲利普斯实验室工作了 4 年。2008 年，她回到隆德大学，开始研究鸟类的磁场定位、导航行为及其生理机制。她研究光依赖磁感应现象的生物物理特性、磁罗盘定位的功能特征，以及磁罗盘与其他罗盘系统的相互作用，特别是偏振光罗盘系统。

图书在版编目（CIP）数据

冈特生态童书.第七辑：全36册：汉英对照 /
（比）冈特·鲍利著；（哥伦）凯瑟琳娜·巴赫绘；
何家振等译.—上海：上海远东出版社，2020
ISBN 978-7-5476-1671-0

Ⅰ.①冈…Ⅱ.①冈…②凯…③何…Ⅲ.①生态
环境－环境保护－儿童读物—汉英Ⅳ.①X171.1-49

中国版本图书馆CIP数据核字（2020）第236911号

策　　划　张　蓉
责任编辑　祁东城
封面设计　魏　来李　廉

冈特生态童书

你迷路了吗？

[比]冈特·鲍利　著
[哥伦]凯瑟琳娜·巴赫　绘

李原原　译

记得要和身边的小朋友分享环保知识哦！
八喜冰淇淋祝你成为环保小使者！